WITHDRAWN

Wildlife Watching

Tide Pool Life Watching

by Diane Bair and Pamela Wright

Consultant:
Eugene P. Williamson
Marine Education Consultant
National Marine Educators' Association

CAPSTONE BOOKS
an imprint of Capstone Press
Mankato, Minnesota

Capstone Books are published by Capstone Press
P.O. Box 669, 151 Good Counsel Drive, Mankato, Minnesota 56002
http://www.capstone-press.com

Copyright © 2000 Capstone Press. All rights reserved.
No part of this book may be reproduced without written permission from the publisher. The publisher takes no responsibility for the use of any of the materials or methods described in this book, nor for the products thereof.
Printed in the United States of America.

Library of Congress Cataloging-in-Publication Data
Bair, Diane.
　Tide pool life watching/Diane Bair and Pamela Wright.
　p. cm.—(Wildlife watching)
　Includes bibliographical references and index.
　Summary: Describes some of the different species of plants and animals found in tide pools and how to go about observing them.
　ISBN 0-7368-0324-6
　1. Tide pool animals—Juvenile literature. 2. Tide pool plants—Juvenile literature. 3. Wildlife watching—Juvenile literature. [1. Tide pool plants. 2. Tide pool animals. 3. Nature study.] I. Wright, Pamela, 1953– . II. Title. III. Series: Bair, Diane. Wildlife watching.
QL122.2.B34 2000
591.769'9—dc21　　　　　　　　　　　　　　　　　　　　　　　　　　99-25189
　　　　　　　　　　　　　　　　　　　　　　　　　　　　　　　　　　　　　CIP

Editorial Credits
Carrie A. Braulick, editor; Steve Christensen, cover designer and illustrator;
　Heidi Schoof, photo researcher

Photo Credits
Brian Parker/TOM STACK & ASSOCIATES, 37
David F. Clobes, 10
Doug Perrine/Innerspace Visions, 40 (top)
Index Stock Imagery/Inga Spence, 13, cover inset; Wayne & Karen Brown, 32
James P. Rowan, 18, 24, 42 (top)
Jay Ireland & Georgienne Bradley, cover, 7
J. Lotter Gurling/TOM STACK & ASSOCIATES, 8
Milton Rand/TOM STACK & ASSOCIATES, 31
Randy Morse/TOM STACK & ASSOCIATES, 17
Thomas Kitchin/TOM STACK & ASSOCIATES, 40 (bottom)
Unicorn Stock Photos/R. Baum, 22; Dennis Thompson, 14
Visuals Unlimited, 41 (top) ; Visuals Unlimited/Glenn M. Oliver, 4; Daniel W.
　Gotshall, 21; Dick Keen, 28; James R. McCullagh, 34; Rick Poley, 41 (bottom); William C. Jorgensen, 42 (bottom); Paul Averbach, 43 (top); John D. Cunningham, 43 (bottom)

Thank you to Steve Miller, Seacoast Science Center, for his assistance in preparing this book.

Table of Contents

Chapter 1   Getting to Know Tide Pool Life.... 5
Chapter 2   Preparing for Your Adventure 11
Chapter 3   Where to Look............................. 23
Chapter 4   Making Observations.................. 29

Features

North American Field Guide.......................... 40
Words to Know 44
To Learn More 45
Useful Addresses 46
Internet Sites 47
Index 48

Chapter 1

Getting to Know Tide Pool Life

Tide pools are pockets of ocean water left on coasts when the tide goes out. Some tide pools are small puddles in the sand or mud. Others are large, deep pools of water between rocks.

Many people enjoy exploring tide pools. They look for animals and plants that live in these pools. You also can learn about tide pools and how to explore them.

About Tide Pool Life

Most tide pool plants are called algae. Algae grow in water or damp places. Algae do not have roots, stems, leaves, or flowers. Algae include

Many tide pools are on rocky coasts.

rockweed, sea lettuce, and Irish moss. Many algae attach themselves to rocks.

Different types of animals live in tide pools. These animals include snails, barnacles, sand dollars, and sea stars. Some tide pool animals attach themselves to rocks or shells. Others swim in the water.

Not all ocean life can live in tide pools. Tide pool life must endure different conditions from plants and animals that live in deep ocean water. Tide pool plants and animals must survive when large, powerful waves crash into the coast. They must be able to live in very warm water. The ocean contains a large amount of water. This water spreads out the sun's warmth. But tide pools contain a small amount of water. The sun quickly heats these smaller bodies of water.

Tide pool algae and animals sometimes must live in dry conditions. The sun can dry up small tide pools. Some tide pool animals close their shells tightly during these times. This traps water inside the shells so the animals can survive.

Sand dollars often live in tide pools.

High and Low Tide

The pull of the moon and sun on ocean water causes tides. This pull is called gravity. The moon and sun pull ocean water toward them.

High tide occurs when ocean water reaches its highest point on the coast. High tide happens at the same time on opposite sides of the earth. High tide usually occurs when the

7

moon is directly over the earth. Water covers most tide pool life during high tide.

Ocean water reaches its lowest point on the coast during low tide. Tide pool life is left on land or in tide pools at this time. Low tide is the best time to look for tide pool life. You can easily observe the algae and animals in tide pools during low tide. You may want to look for tide pool life during a new or full moon. The tides are especially low during these times.

High and low tide usually happen twice each day at different times. People can predict when these times will occur. Sporting supply stores near ocean coasts sometimes have lists of these times. Most coastal city newspapers also print high and low tide times.

The best time to see tide pool life is during low tide.

Chapter 2

Preparing for Your Adventure

Learn about tide pool life before you visit tide pools. Check out books about tide pools from your school or local library. Study pictures of tide pool life. This will help you identify plants and animals in tide pools. You may want to study a field guide about tide pool life. Field guides show what animals look like and tell where they live. This book has a short field guide on pages 40 to 43.

Marine science centers and places with large aquariums also may have information about tide pool life. Workers at these places can teach you about tide pool plants and animals. Some of

Check out books about tide pool life from a library.

these places offer tide pool walks. You can observe tide pool life with a guide during these walks.

What to Bring

Bring a plastic bag with a tight seal to view tide pool life. Fill the bag with ocean water. You can place most tide pool animals in the bag. This will allow you to closely observe the animals. Do not place algae in a bag. Many algae are attached to rocks. These algae will die if they are removed from the rocks.

Always return the animals to the same place where you found them. You may harm the animals if you place them in a different location.

Do not bring a pail to view tide pool life. Tide pool plants and animals are difficult to observe in pails.

You may want to bring a magnifying glass. These small, hand-held lenses make objects appear larger. You will be able to observe tide pool life closely with a magnifying glass.

Guides may take groups of people on tide pool walks.

Bring materials to record information about tide pool life. These materials may include a notebook, pencils, and waterproof markers. You may want to bring a waterproof camera to take photographs. Water can ruin cameras that are not waterproof. Be sure to put your recording materials in waterproof bags.

Dress properly for the weather when you watch tide pool life.

Bring a backpack to carry all of your materials. A backpack will leave your hands free to explore tide pools. It is important to have your hands free in case you slip on rocks.

What to Wear

Wear rubber-soled shoes or boots when you visit tide pools. These shoes or boots will help you grip slippery areas. They also will protect your feet from sharp shells and rocks. Do not wear sandals. Sandals usually do not grip surfaces very well. They may slip off your feet. Wear rubber gloves to protect your hands from sharp or stinging plants and animals. Rubber gloves also can help keep your hands warm.

Dress properly for the weather and season. Wear shorts and short-sleeved shirts in warm weather. But remember to bring a jacket. The weather may become cooler. You may want to wear layers of clothes. This will allow you to put clothes on or take them off as the weather changes. Bring extra clothes and a towel. You may want to change clothes and dry yourself if you get wet.

Dress warmly if you watch tide pool life during cold weather. Wear a heavy jacket. Bring an extra pair of gloves to keep your hands warm. Wear a hat or hood to keep your head and ears warm. Wear heavy socks and warm boots to protect your feet from the cold.

Protect yourself from the sun when you watch tide pool life. Wear sunscreen to protect your skin from sunburn. Wear sunglasses or a hat to protect your eyes from the sun. Sunglasses with polarizing lenses work well to watch tide pool life. Polarizing lenses reduce glare from the sun on the water. They can help you see better as you look into the water.

Make sure you bring all supplies back home and place all trash in garbage containers. This helps protect the environment from pollution. Pollution can harm ocean plants and animals. For example, plastic bags in oceans can harm sea turtles. Sea turtles sometimes eat these bags and die.

Safety

Be careful when you explore tide pools. Children should watch tide pool life with an adult. Adults can help if children become hurt or lost. Watch where you step in tide pools. The rocks in tide pools often are slippery. Walk on rocks that are close to the coast. Do

Algae make rocks slippery.

not walk near large waves. Large waves may knock you down or pull you into deep water.

Avoid exploring tide pools near water with undertows. These strong currents are below the surface of the water. They usually flow in the opposite direction of the surface water. Undertows can pull you into deep water. You can ask a guide if there are any places with undertows if you watch tide pool life at a park. You may ask a lifeguard if you watch tide pool life at a beach.

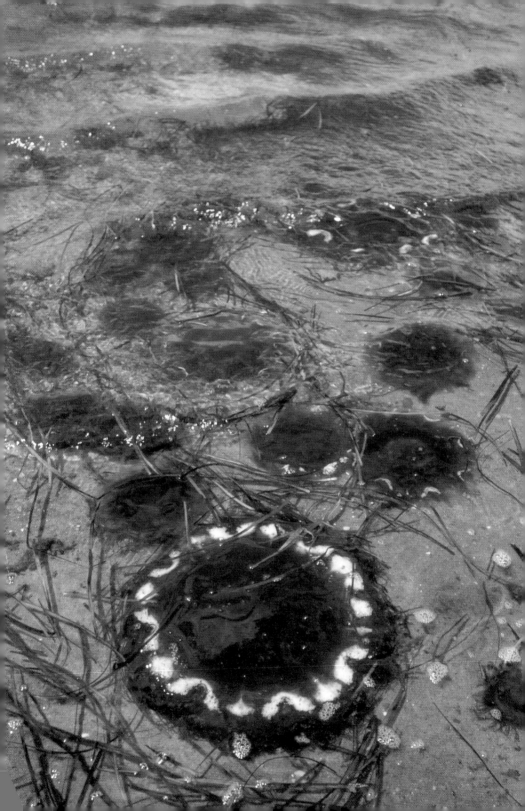

Make sure to check the times of low and high tides. These times change every day. Look for tide pool life during low tide. Leave tide pool areas before high tide.

Stay on marked beach paths or boardwalks when you walk to and from tide pools. Boardwalks are walkways made of thick boards on beaches. You may destroy beach plants if you wander away from marked walking areas. Beach plants are important to the environment. They help control beach erosion. Beach erosion occurs when water wears away beach sand.

Handling Tide Pool Life
You may want to handle some tide pool life. Handle these plants and animals gently. Rough handling may harm them. Do not drop tide pool animals. Do not keep them out of the water for more than a few minutes. Never pull off algae or animals that are firmly attached to rocks or other objects. This can injure them.

Some animals are dangerous to touch. Always wear gloves to protect your hands. Never touch

---

Jellyfish may sting if you touch them.

jellyfish. Some jellyfish will sting if you touch them. These stings can hurt you. They may cause blisters or rashes. Even dead jellyfish can be dangerous. Do not handle sea anemones. Sea anemones have tentacles. They use these limb-like parts to sting other animals for food. They may sting you if you touch them. Do not handle crabs. Crabs have sharp front claws called pincers. They may use their pincers to snap at your fingers. Some worms also sting or bite.

Leave all tide pool animals, shells, and other items you find on the coast. Most tide pool animals cannot survive outside of their natural environments. Some tide pool animals use shells for shelter. These animals may not survive if you remove shells from the coast.

Make sure you put rocks and shells back exactly where you found them. Return rocks and shells to their original positions. Plants and animals may live on a certain side of rocks and shells. They may only be able to survive on this side.

Sea anemones have many tentacles.

Chapter 3

Where to Look

Visit ocean coasts to see tide pool life. There are four oceans in the world. These are the Pacific, Atlantic, Indian, and Arctic Oceans. The Atlantic and Pacific Oceans border North America. The Atlantic Ocean borders the continent to the east. The Pacific Ocean is west of North America.

Intertidal Zone

People divide an ocean into many zones. The intertidal zone is the area of an ocean's coast the tides cover and uncover.

You can look for tide pool algae and animals in different parts of the intertidal zone. Some algae and animals live in the upper part

Tide pool algae and animals live in the intertidal zone.

of the intertidal zone. This part is located the greatest distance from the ocean. These plants and animals live on land most of the time. Plants and animals in the middle part of the intertidal zone live both on land and in water. Tide pool life in the lower level of the intertidal zone is almost always in water.

Habitats

Look for tide pool life in their habitats. These are the natural places and conditions in which plants and animals live.

Different tide pool algae and animals live in different habitats. Many tide pool algae and animals live along rocky coasts. Tide pool animals such as mussels and barnacles cling to rocks. This helps them survive the force of ocean waves. Other tide pool animals have different habitats. Hermit crabs live in shells. Clams and sand dollars bury themselves in the sand.

Barnacles cling to rocks in tide pools.

Places to See Tide Pool Life

1 **Acadia National Park, near Bar Harbor, Maine:**
Acadia National Park is located on Mount Desert Island. Tide pool walks are available. Visitors can view a variety of tide pool life in this park.

2 **Odiorne Point State Park, Rye, New Hampshire:**
The Seacoast Science Center is located in this park. The center offers many tide pool walks and programs on the park's beaches. Visitors can see tide pool animals such as sea stars, sea anemones, and periwinkle snails.

3 **Shelter Cove, Humboldt County, California:**
Shelter Cove is a small town located on a remote area of land called the Lost Coast. Visitors can view a variety of tide pool life on beaches near the town.

4 **Natural Bridges State Park, Santa Cruz, California:**
This park is located on the west side of Santa Cruz. Visitors can watch tide pool life on the park's beach. Tide pool walks occasionally are available.

5 **Deception Pass State Park, Oak Harbor, Washington:**
This popular park is located on Whidbey Island. Tide pool walks often are conducted here. Visitors can see a variety of tide pool life such as mottled sea anemones and various types of algae.

6 **Yaquina Head Outstanding Natural Area, near Newport, Oregon:**
This area is located on a narrow piece of land in the Pacific Ocean. Visitors can view tide pool animals such as purple sea urchins, sea anemones, barnacles, turban snails, and hermit crabs. Yaquina Head Outstanding Natural Area has tide pools that people who are disabled can visit. The Oregon Coast Aquarium and the Hatfield Marine Science Center also are located near this area. Visitors can see indoor tide pool and marine displays at these places.

7 **Ecola State Park, near Cannon Beach, Oregon:**
This park includes 9 miles (14 kilometers) of Oregon's coast. Visitors can see a variety of tide pool life on Indian Beach.

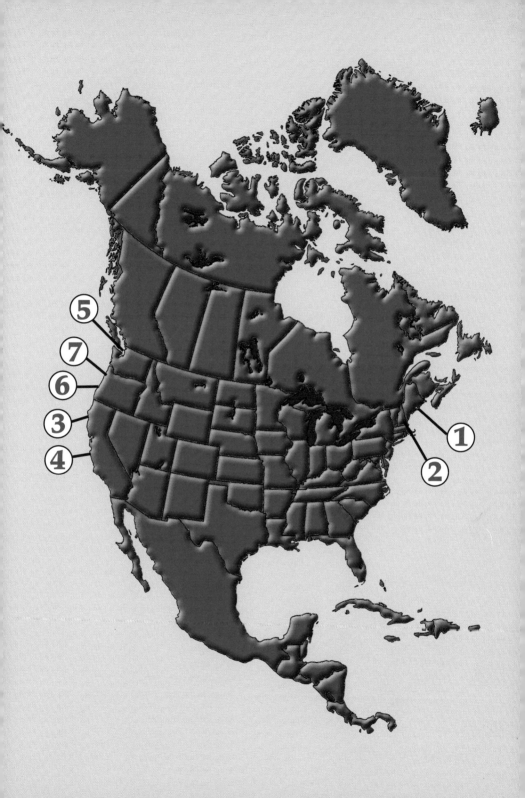

Chapter 4

Making Observations

You can observe different features and behaviors of tide pool life. You may observe their colors, shapes, and sizes. You may watch some tide pool animals eat. You then can record your observations about tide pool life.

Algae

Three groups of algae live in tide pools. These are green, red, and brown algae. Green algae are the most common. Algae in these groups are not always the same colors as the group names. For example, red algae are not always red. They may be purple or orange.

Rangers may teach you how to properly observe tide pool life at some places.

Sea lettuce is one common green alga. You may find sea lettuce in tide pools. Sea lettuce is bright green and leafy. It may grow as long as 3 feet (.9 meter). Look for sea lettuce on rocks or sand near the water.

Irish moss is a red alga. Look for patches of this dark red alga on tide pool rocks. Irish moss usually grows 2 to 4 inches (5 to 10 centimeters) long. You may see many small, flat branches if you look at Irish moss with a magnifying glass.

Rockweed is a common brown alga with flat branches. Rockweed is covered with bumps called air bladders. Air bladders cause rockweed to float up to the tide pool's surface. This alga uses the sunlight it receives on the water's surface to grow. Look for large groups of rockweed on rocks in the water. You may see piles of rockweed along the coast.

Mollusks

You may find mollusks in tide pools. Mollusks have soft bodies. Many are protected by hard, outer shells.

Rockweeds are covered with air bladders.

The mollusk group includes snails. Snails cling to rocks or bury themselves under the sand in tide pools. Most snails have a single, spiral-shaped shell. It may be difficult to see a snail's head. A snail often will pull its head into its shell if it senses you. Each snail has a foot under its shell. A snail uses its foot to crawl on rocks and sand.

Clams are mollusks that live underneath the sand in tide pools. Clams have a top and a

Many sea anemones are brightly colored.

bottom shell. These shells are hinged together. Each clam has muscles to hold its shell closed. A clam has a soft body with no head inside its shell. Most clams have one foot. A clam uses its foot to bury itself in the sand. It sticks its foot into the sand. The clam then uses muscles in the foot to pull its body into the sand.

Cnidaria

Some tide pool animals are members of a group called cnidaria (nye-DARE-ee-ah). Cnidarians have hollow bodies. Tentacles surround their mouths. These tentacles have special cells filled with poison. Most cnidarians use their tentacles to sting plankton. Plankton are tiny plants and animals that float in ocean water. The tentacles release poison that paralyzes the animals. Cnidarians then push the poisoned animals into their mouths.

Sea anemones belong to the cnidaria group. Sea anemones often cling to rocks. They look like flowers. Some are brightly colored. They may be pink, purple, green, or red.

You may not be able to observe sea anemones' tentacles. Sea anemones close up during low tide. This allows them to hold water to survive.

Jellyfish also are cnidarians. These umbrella-shaped animals float in the water. They sting and eat plankton and other small ocean animals that come near them.

Most sea stars have five arms.

Anthropods

Some tide pool animals are anthropods. These animals have soft bodies. A strong material called chitin covers their bodies. The anthropod group includes insects.

Crabs are anthropods. Crabs often bury themselves in sand. Crabs have long, thin legs. They have pincers attached to their legs. Crabs use their pincers and legs on one side of their bodies to push themselves sideways.

Barnacles also are anthropods. Barnacles live in groups called colonies. Most barnacles are gray-white and less than 1 inch (2.5 centimeters) long. Barnacles fasten themselves to rocks and the shells of other animals. Barnacles also cling to the bottoms of boats. They live on the same objects their entire lives. Do not lift barnacles away from surfaces. This will kill them. Use a magnifying glass to help you observe barnacles.

Amphipods are small anthropods that look like insects. You may see amphipods swimming in tide pools. You also may see amphipods crawling on sand. Most amphipods are less than 1 inch (2.5 centimeters) long. They are flat and have arched backs. Most amphipods have seven pairs of legs. Some amphipods have legs on their tails. They use these legs to jump.

Echinoderms

You may find echinoderms (eh-KEE-noh-durms) in tide pools. These animals have hundreds of hollow feet with suction cups.

Sea stars are echinoderms. Sea stars often are called starfish. Look for sea stars at the bottom of tide pools. They have several arms that look similar to the points of a star. Most sea stars have five arms. Some sea stars have more than 15 arms. You may find sea stars with one or two arms missing. These sea stars will grow back new arms.

Sea stars cannot live out of the water for more than a few minutes. Put sea stars in your bag of ocean water if you want to closely observe them. Use a magnifying glass to look at the tips of sea stars' arms. You may notice tiny red dots on some sea stars. These are the sea stars' eyes.

Sea urchins also are echinoderms. Sea urchins live on rocks or in sand at the bottom of tide pools. You might find sea urchins under rocks or algae. Sea urchins look like small balls covered with spikes. The spikes are not sharp. A sea urchin's mouth is on the bottom of its body. It eats as it moves along rocks. Sea urchins sometimes eat small snails and barnacles.

You may put sea urchins in your bag of water for a closer look. Do not view sea urchins out of water. They cannot survive out of water for long periods of time.

Sea urchins are covered with spikes.

Worms

Many worms live in tide pools. Look for worms under rocks. You may find them under shells or algae. Many worms live in the sand.

Worms may have a variety of features. They can be one of several colors. Worms may be brown or gray. Some are pink, green, or blue. Some worms are covered with spines. Others have tentacles.

Make an Algae Mounting

You may want to make an algae mounting. This will allow you to keep track of the different types of algae you find.

1. Collect small pieces of algae from the shore. Do not take a large amount of the algae. Some animals need algae for food. Do not take living algae off rocks or other surfaces. This can damage or kill the algae. You may want to gather algae that waves have brought onto the shore. Do not gather kelp, rockweed, or other coarse algae to mount. These algae often shrink and do not stick well to paper.

2. Rinse dirt off the algae. Place the algae on a piece of herbarium paper or heavy drawing paper. People use herbarium paper to dry and mount plants. You can purchase it at some garden centers.

3. Put the paper and algae in a tray filled with water. It is best to use ocean water. Arrange the algae on the paper while they are under water.

4. Lift the algae and paper from the tray. Make sure the algae keep their shape. Hold the paper up to drain the extra water.

5. Lay the algae and paper face up on a piece of newspaper. Cover the algae with a piece of fine cloth. You can use pieces of old sheets or pillow cases. Place weight on top of the cloth. You may have more pieces of algae you want to mount. You can place more layers of newspaper, algae, and cloth on top of the first one to mount these algae.

6. Replace the newspaper after about two hours. Then replace the newspaper once or twice daily. Wait until the algae are dry. Some algae dry in a few days. Other algae may take about two weeks to dry. Make sure the algae are completely dry before you remove them from the layer. Algae can shrink if they are not dry. This can cause the paper to wrinkle or curl up.

7. Some algae stick firmly to the paper after they are dry. You may fasten the algae to the paper with glue if they do not stick. Do not use rubber cement. Rubber cement can cause the paper to wrinkle.

8. Label the algae mounting. Write down the name of the algae, where you found them, and the date.

Recording Your Observations

You may want to keep records of the tide pool life you observe. Make a list of the plants and animals you see in tide pools. Note where you found them. Write down the time of day. Note tide pool animals' behavior. You may want to make drawings of plants and animals you see. You also may take photographs of tide pool life.

Record information each time you go tide pool life watching. This will help you keep track of your observations. It also will help you remember your tide pool adventures.

North American Field Guide

Cushion Sea Star

Description: Cushion sea stars are the largest sea stars that live on the Atlantic coast. Cushion stars can be red, orange, or yellow. They may be 10 inches (25 centimeters) wide. Cushion sea stars have five arms. They have fat bodies with ridges running across them.

Habitat: On sand at the bottom of tide pools

Food: Clams, barnacles, snails, oysters

Pink-Tipped Sea Anemone

Description: Pink-tipped sea anemones live on the Pacific and Atlantic coasts. These sea anemones are white. The tips of their tentacles are bright pink. Pink-tipped sea anemones may be green if algae live in their tissues. Pink-tipped sea anemones may grow up to 1 foot (.3 meter) wide. These anemones usually grow in large groups.

Habitat: Rocky areas

Food: Plankton, small fish

North American Field Guide

Sea Lettuce

Description: Sea lettuce is a bright green alga. Sea lettuce forms flat sheets on coasts. These sheets usually are about 6 inches (15 centimeters) to 1 foot (.3 meter) long. Waves sometimes wash sea lettuce onto the coast. It can live through large temperature differences. Some people eat sea lettuce.

Habitat: Rocks, shallow water

Fiddler Crab

Description: Many fiddler crabs live on the Atlantic coast of North America. Some fiddler crabs are gray or black. Most fiddler crabs are about 1 to 2 inches (2.5 to 5 centimeters) long. Male fiddler crabs have one claw that is much larger than the other. They use this claw to defend themselves from enemies and to attract mates.

Habitat: Sandy and muddy areas on land, under sand during high tide

Food: Bacteria in sand, dead plants and animals

North American Field Guide

Hermit Crab

Description: Hermit crabs are found in tide pools throughout the world. Some hermit crabs are dark red. They may have orange-red legs. Hermit crabs sometimes are called false crabs. They have three pairs of walking legs. Most other crabs have four pairs of walking legs. Hermit crabs also have two claws. They use their larger left claw to protect themselves from enemies. They cover the opening to their shells with this claw when they sense danger. Hermit crabs are nocturnal. Nocturnal animals are most active at night. Hermit crabs live in snail shells to protect their soft abdomen. They often live near other hermit crabs.

Habitat: Inside snail shells on land

Food: Dead plants and animals

Talitrid Amphipod

Description: Talitrid amphipods can be found on the Pacific and Atlantic coasts. Talitrid amphipods are about .3 inch (.8 centimeter) long. They sometimes are called beach hoppers. Some are pale orange. Others are gray. Talitrid amphipods are nocturnal. They move to the lower intertidal zone to hunt for food.

Habitat: On coarse sand in the upper intertidal zone

Food: Dead plants and animals

North American Field Guide

Bristle Worm

Description: Bristle worms are segmented worms. They are covered with spines. Most bristle worms are less than 1 inch (2.5 centimeters) long. Bristle worms have sharp hooks around their mouths. They use these hooks to catch other animals.

Habitat: Under sand and rocks

Food: Amphipods, other worms

Periwinkle Snail

Description: Many periwinkle snails live on the coasts of North America. Some periwinkles are tan, brown, gray, or black. They may have spots or stripes. Most periwinkles are about 1 inch (2.5 centimeters) long. These snails release a slimy covering over their bodies when the tide goes out. This prevents them from drying up.

Habitat: On rocks

Food: Algae on rocks

Words to Know

algae (AL-jee)—small plants that grow in water or on damp surfaces; algae do not have roots, stems, leaves, or flowers.

erosion (i-ROH-zhuhn)—the gradual wearing away of land by water or wind

high tide (HYE TIDE)—the highest level the tide reaches on the coast

intertidal zone (in-tur-TYE-dul ZOHN)—the area of an ocean's coast that is underwater at high tide and exposed at low tide

low tide (LOH TIDE)—the lowest level the tide reaches on the coast

pincer (PIN-sur)—a crab's claw

plankton (PLANGK-tuhn)—tiny plants and animals that float in the ocean

tentacle (TEN-tuh-kuhl)—a long, flexible limb of some ocean animals; animals use tentacles to touch, grab, or smell.

To Learn More

Barnhart, Diana and Vicki Leon. *Tidepools: Bright World of the Rocky Shoreline.* Close Up. Parsippany, N.J.: Silver Burdett Ginn, 1995.

Gunzi, Christiane. *Tide Pool.* Look Closer. New York: DK Publishing, 1998.

Hansen, Judith. *Seashells in My Pocket: A Child's Nature Guide to Exploring the Atlantic Coast.* Boston: Appalachian Mountain Club Books, 1992.

Kricher, John. *Peterson First Guide to Seashores.* Boston: Houghton Mifflin, 1992.

Tibbitts, Christiane Kump. *Seashells, Crabs, and Sea Stars.* Young Naturalist Field Guides. Milwaukee: Gareth Stevens, 1998.

Useful Addresses

Alaska SeaLife Center
301 Railway Avenue
P.O. Box 1329
Seward, AK 99664

**Department of Fisheries and
 Oceans–Canada**
Communications Branch
200 Kent Street
13th Floor, Station 13228
Ottawa, ON K1A 0E6
Canada

National Ocean Service
National Oceanic and Atmospheric
 Administration
SSMC4, 13th floor
1305 East West Highway
Silver Spring, MD 20910

Internet Sites

Alaska Department of Fish and Game—Intertidal Animals
http://www.kodiak.org/tidal.html

Cabrillo High School Aquarium
http://www.cabrillo-aquarium.org

Life in a Tide Pool
http://www.columbiafalls.sad37.k12.me.us/tidepool/Intro.html

Monterey Bay Aquarium
http://www.mbayaq.org/index.htm

Seacoast Science Center
http://www.seacentr.org

Index

alga, 5–6, 9, 12, 19, 23, 25, 29, 30, 36, 37, 38
amphipods, 35

barnacles, 6, 25, 35, 36

camera, 13
clam, 25, 31–32
clothes, 15
coast, 5–7, 9, 16, 20, 23, 25, 30
crabs, 20, 25, 34

field guide, 11

gravity, 7

high tide, 7, 9, 19

intertidal zone, 23, 25
Irish moss, 6, 30

jellyfish, 20, 33

low tide, 7, 9, 19, 33

magnifying glass, 12, 30, 35, 36

pincers, 20, 34
polarizing lenses, 16

rockweed, 6, 30, 38

sand dollars, 6, 25
sea anemones, 20, 33
sea lettuce, 6, 30
sea stars, 6, 35–36
sea urchins, 36
shell, 6, 15, 20, 25, 30, 31–32, 35, 37
snail, 6, 31, 36

tentacles, 20, 33, 37

waves, 6, 17, 25, 38
worms, 20, 37